Dedicated to our children, Marissa and Mark,
who helped us dream new dreams.

Special thanks to all the lavender farms who shared their stories,
and the photographers who captured the beauty of the fields.

ISBN 978-1-7329699-3-3

Cover Photo: J.P. Dobrin

Printed in the United States of America
Lion Heart Publishing
8537 Sonoma Highway, Kenwood, CA 95452
lionheartpub@gmail.com

LAVENDER FIELDS
of AMERICA
A NEW CROP OF AMERICAN FARMERS

Rebecca Rosenberg and Gary Rosenberg

Some odors wrinkle the nose, and some turn down the mouth, and some cause a sharp inhale. But the scent of lavender lifts the nose. It's instinctive, as if the body demands to know where the scent is coming from. When a breeze carries an unexpected trace of lavender, the head tilts back and the nose comes up, seeking its source.

Lavender's power may be because scents are instantly evocative. Poet Diane Ackerman says, "One scent can be unexpected, momentary and fleeting, yet conjure up a childhood summer beside a lake in the mountains." Lavender may bring back a friend from long ago, a hike through a farm field, or an enchanted evening. The smell can unlock potent memories from the past.

But what, precisely, does lavender smell like? Can we get closer than to say that lavender smells like purple, or that lavender smells like Cousin Abigail, or that lavender simply smells wonderful?

Perhaps its scent is like humor: to define it is to ruin it. Or maybe lavender's aroma is too grand to define; smelling it is like "inhaling the earth," to use Czech poet Rainier Maria Rilke's term. Different varieties of lavender have distinct fragrances, but here is a rough attempt at putting into words what lavender smells like: add together some sage, mint, smoke, and cinnamon, and then ask Cousin Abigail to stir it with her finger, and you'll produce the scent of lavender. Or maybe not quite.

—Novelist James Thayer

CONTENTS

SONOMA LAVENDER PHOTO BY REBECCA GOSSELIN

The American map of red and blue states is blurring into a blissful state of lavender. We are not talking politics. There are more than 250 lavender farms across America, growing the sweet-smelling crop which England and France made famous.

Why lavender? For centuries, people have known that lavender has solved a vast number of maladies, from relieving stress to repelling insects and even healing wounds. Now modern research has proven that lavender's healing powers are more than folklore and imagination.

What is it about lavender that has drawn people to change their lives and stake their livelihood on this flowering herb? In the past two decades, lavender farms have sprung up across the country from Hawaii to Maine, including the authors' own farm, Sonoma Lavender in Kenwood, California. *Lavender Fields of America* explores the reasons behind this new crop of American farmers and the magic of the herb that transformed their lives. If you haven't experienced a lavender field in full bloom, put it on your must-do list. Discover the beauty, peace and healing power of lavender.

ALI'I KULA LAVENDER

MAUI · HAWAII

A single lavender plant, a gift from a dear friend, started a journey Ali'i Chang would never regret. As the morning mist lifted on the Maui upcountry hills, he envisioned rows of the purple herb where they had never grown before. Naysayers tried to discourage him, citing the arid climes of Provence. Ali'i smiled quietly and planned his gardens.

Agricultural artist and horticultural master, Ali'i Chang (1942-2011), wanted more than lavender fields. He wanted a garden dedicated to "Aloha", the spirit of friendship and love with each other and the earth: a place where people could experience Maui's Aloha spirit in a new way. He created a destination which blossomed into a work of art and his legacy continues on.

Nestled along the skirt of Haleakala (House of the Sun) volcano, Ali'i Kula Lavender resides in the rustic town of Kula at an elevation of 4,000 feet. Walking paths wind around manicured fields of lavender. The air is crisp and clean. Wind chimes ring in the cool breeze. Fountains trickle with water. Statues of Buddha and Quan Yin, the Buddhist Goddess of Compassion, create an other-worldliness. Rosemary and camelias add to the zen fragrance of the surrounding lavender. At the top of the hill, a gazebo invites contemplation and rest. A breathtaking 180 degree view extends from the apex of the Haleakala volcano, down to lower Maui, and across the Pacific to the islands of Kaho'olawe, Lana'i and Molokai.

"We are not separate from the earth, we are a part of it."

–Pomai Weigert, Ali'i Kula Lavender

Ali'i Kula is the first and only lavender farm in Hawaii with more than 55,000 lavender plants. Over forty-five different varieties of lavender bloom on thirteen and a half acres. Although lavender is not native to Maui, it has settled on the majestic mountain with style and grace in Kula's dry climate, blooming year round, with peak color June through August. Hundreds of visitors come to the unique lavender farm daily, taking advantage of cart and guided walking tours, garden and craft classes and special events.

Ali'i Kula Lavender works with local artisans to create lavender products which promote a health and wellness lifestyle and in turn creates a sustainable economy in Maui through agri-tourism. Their products enrich the quality of life through rejuvenation, comfort and serenity, and are available at their gift store and website.

Their dedication to the Aloha spirit influences everything from sustainable farming practices to involvement in the community. Ali'i Kula Lavender believes in learning from those who came before us through culture and values. "We respect the earth and the environment by using earth friendly materials and organic ingredients as much as possible," says Pomai Weigert, Marketing and Community Relations.

http://www.aklmaui.com

**Signature product:
Lavender Lilikoi Jelly**

Lilikoi is a delicious passion fruit grown in Upcountry, Maui. Infused with lavender, it creates a tangy, sweet jelly that is excellent on biscuits, toast, scones, bagels, or as a glaze over Cornish hens or ham. Add two scoops in a pan when frying shrimp and create your own version of sweet & sour shrimp.

Lavender Gourmet Seasoning Crusted Lilikoi Shrimp
by Chef Crystal Carroll

12 raw shrimp, washed, shelled and de-veined
Lavender Gourmet Seasoning (A dry rub of Lavender, Hawaiian Salts and Kukui nut with other herbs and spices)
1 tsp. canola or olive oil
3 tbsp. Lavender Lilikoi Jelly
1/4 head cabbage (mix of Chinese, red, and regular cabbage is best)
Italian salad dressing
1 T. lavender buds

Cover shrimp lightly with *Lavender Gourmet Seasoning*; let sit for 15 minutes.

In a sauce pan big enough to hold all the shrimp, add oil and Lilikoi Jelly. Cook shrimp on medium-high heat until shrimp turns pink.

To complete the presentation, finely cut cabbage and mix with a crushed table-spoon of lavender and Italian dressing.

Place cabbage on a plate and top with shrimp.

11

MT. SHASTA LAVENDER FARMS

MONTAGUE • CALIFORNIA

"Yes, we do work 24/7 four to five months of the year. Fortunately lavender is a rewarding plant," says Gail Winslow, Ph.D.

Gail and her husband, David McGee-Williams, Ph.D., both practicing psychologists, turned to gardening to counter their sedentary jobs. Gardening became a passion and soon they ran out of space in their one-acre property in town. They needed to expand, stretch out. With an eye toward a retirement project, they set out on a stifling day in August 2000 to explore an 800 acre parcel, with their two daughters.

The real estate agent pointed out the unforgiving terrain; impossibly rough and rocky. There was no water, no road, no power, and no buildings on this formidable property. But the awe-inspiring view of Mt. Shasta, rising 14, 000 feet above sea level , made it all seem worthwhile. The family voted unanimously to purchase the property against advice from real estate agents.

The couple considered planting a rose garden but threw the idea aside in favor of lavender. A hedge of lavender had survived ten winters under six feet of snow in their yard. Why not start a lavender farm? They began reading every book they could find on lavender and traveled to France to see lavender farms firsthand.

Back home, Gail and David continued their work as psychologists while

PHOTO: LANI PHILLIPS

PHOTO: GAIL WINSLOW

PHOTO: DAVID McGEE-WILLIAMS

PHOTO: ROXI MUELLER

"The farm brings us so many joys. We enjoy the mountain sunsets, the exquisite privacy, the immediacy of the weather, caring for the plants that now have us thoroughly seduced."

–Gail Winslow, Mt. Shasta Lavender Farms

making progress on the property in their spare time. Blissfully unaware of the enormity of the undertaking ahead, the couple recalled childhood memories of grandparents' farms, which didn't include backbreaking labor, long hours, and the caprice of nature. They ploughed in a road and drilled a well. Water proved plentiful. With the help of energetic friends and family, they cleared the land and prepared for planting.

In 2001, they planted 4000 plants; *Lavendula x intermedia* 'Grosso' and *Lavendula angustifolia* 'Buena Vista.' By 2003, the field opened to the public. Through many more years of great discouragement, intense labor, utter exhaustion and unexpected expense, Mt. Shasta Lavender Farm now has 55,000 plants, more than eleven acres, with plans to expand. The beauty and rewards of the plant and the enthusiasm from Shasta Valley neighbors and visitors to the farm have kept Gail and David highly motivated. The family now lives at the farm, making a longer commute for Gail and David, who maintain their practices.

In 2006, they created a lavender labyrinth designed in the pattern of the Chartres Cathedral in France. It is planted in *L. angustifolia* 'Hidcote,' a sweet smelling inspiration for those who follow the eighteen-minute meditative walk.

Mt. Shasta Lavender Farms distills its own oil and cures it for a year. They use it in their lavender products, which are sold in a stone cottage built for that purpose. The farm is open every day for seven weeks in summer.

After ten years, longer than it took to get their graduate degrees, they feel they have truly become farmers. They enjoy nurturing the lavender plants that seem to lift the spirits and bring a sense of total relaxation to the thousands of visitors who stop by each year.

http://www.mtshastalavenderfarms.com

The ambient odor of lavender reduces anxiety and improves mood.

A recent medical study proved that lavender reduces anxiety and improves one's mood. Heady findings. The goal of this study was to investigate the impact of lavender on anxiety, mood, alertness and calmness in dental patients. Two hundred patients were stimulated with the ambient odor of lavender. These conditions were compared to a music condition and a control condition (no odor, no music).

Anxiety, mood, alertness and calmness were assessed while patients waited for dental treatment. Statistical analyses revealed that lavender reduced anxiety and improved mood in patients waiting for dental treatment. These findings support the previous opinion that odors are capable of altering emotional states and indicate that the use of odors is helpful in reducing anxiety in dental patients.
University Clinic of Neurology, Medical University of Vienna, Austria, Department of Medical Statistics, Medical University of Vienna, Austria, University Witten Herdecke, Germany 10 August 2005.

Reduce anxiety and improve mood with a lavender eye pillow or sleepmask. Or set a lavender diffuser in a breezy place for an uplifting scent of lavender in the air. At the end of a long day, heat a lavender neck pillow for heat and aromatherapy.

SONOMA LAVENDER

KENWOOD · CALIFORNIA

Late one Sunday afternoon in 1996, Gary and Rebecca Rosenberg wandered into a derelict garden in Stinson Beach north of San Francisco. An intoxicating scent drew them to a gnarly shrub, gray branches stretching helter-skelter to the sky. The ancient lavender plant had not been pruned for years.

They rationalized they would help the plant by snapping off brittle branches. They took a bundle home to their bedroom and lay it in a willow basket. Every time they walked into the room the heady scent soothed their spirits.

Life had become too hectic with two small children, two businesses, and two homes. Every weekend the family headed to their weekend home in Sonoma County to escape the frenetic pace of their advertising agency and catalog company. They inched through knotted traffic on the Golden Gate Bridge in Gary's prized 1958 Jaguar, singing oldies with their children, Marissa and Mark, to drown out the whirring engines along the highway. The two hour drive took them far from the fast lane of daily life.

As soon as they turned up the Sonoma Valley, cradled by the Mayacamus Mountains to the east and the Sonoma Mountains to the west, with clear skies and vineyards, a peace-of-mind enveloped the family. Soon they reached their Kenwood property. Their son, Mark, searched the dry

PHOTO: REBECCA GOSSELIN

PHOTO: GARY ROSENBERG

*"Lavender draws me in with its intoxicating scent,
and inspires me to find new ways to show off its powerful benefits."*

– Rebecca Rosenberg, Sonoma Lavender

creek for arrowheads instead of playing computer games. Their daughter ran to pet the resident cows. Gary dug in the garden and painted furniture. Rebecca gazed out at the grassy acres wondering how they could change their crazy lives and stay in Sonoma Valley forever. When the weekend was over the family piled in the car, longing to stay in the country.

Finally, Gary and Rebecca made the decision to sell their businesses and move to Sonoma to create a slower way of life for their family. Rebecca started reading about the benefits of lavender and dreaming of fields of purple.

Gary scoffed. "What can you make with lavender?"

"Lavender is like magic," Rebecca said. "You can make so much. It repels insects. It heals the skin. It relaxes the muscles. It helps you sleep." She bought a fifty pound bag of lavender and started sewing sachets and eye pillows. She made soap and candles in the kitchen. Soon she began selling them to local stores.

Surprised by the strong reception Rebecca was getting with the lavender products, Gary started taking organic farming and herb classes. He knew their fields had a high water table that flooded with spring runoff from Sugarloaf Mountain. He had deep trenches dug and installed French drains to draw off underground water. He surveyed the land with string and chalk, and created twelve inch mounds to keep the lavender roots from being too wet. By May they planted 5,000 lavender plants, *L. angustifolia* 'Hidcote,' and *Lavandins* 'Grosso' and 'Provence.'

23

Lavender to relax muscles.

Lavender combined with heat therapy on achy muscles works wonders. The aromatherapy of lavender relaxes, while heat therapy soothes and heals achy muscles. Heat a lavender neck pillow, heat wrap or spa blankie to soothe achy muscles. Lavender Spa Booties soothe achy feet.

To learn more about lavender, the Rosenbergs traveled to le Pays de Sault in Provence, where it has been grown for centuries. The farmers shared their methods of farming and distillation. Gary and Rebecca sat at a farmer's kitchen table, smelling the various lavender oils, while the Provencal patriarch explained in French the quality and benefits of each. They learned about 'terroir,' the French word meaning 'feeling of the land.' Growing lavender seemed similar to growing wine grapes, taking advantage of the particular soil, sun and temperature.

While French farming was impressive, their lavender sachets and lotions did not seem to use the bountiful benefits of lavender. Rebecca felt there was much more to be created with lavender. She expanded the Sonoma Lavender product line to take full advantage of lavender's magic. She created lavender and flaxseed booties, neck pillows and teddy bears that could be heated in the microwave to soothe achy muscles. They sewed computer keyboard pillows to tame office stress, pillow liners to encourage sleep, and closet hanger covers to repel moths. Today, Sonoma Lavender markets 300 lavender products to more than 4,000 stores across the country and around the world.

"We were drawn to the healing powers of lavender and have shared it with millions of lavender enthusiasts through our products," Rebecca says. "We hear amazing stories of healing."

The farm is open only two days a year for the Sonoma Lavender

PHOTO: ANANDA FIERRO

Festival, when 5,000 people enjoy the lavender fields. Gary leads a tour through the Lavender Specimen Garden, teaching about the varieties of lavender. Demonstrations and classes under the 400 year old oak include lavender cultivation, cooking with lavender and the magic of lavender. Crafters make lavender wreaths and wands. There is a spa tent for lavender massage and a bubble room to experience lavender lotions and potions. Live music, wine and lavender cuisine tempt people to stay longer and longer every year. "There's nothing like sharing a field of lavender in full bloom with people," Gary says. "A look of pure joy floods their faces, young and old."

http://www.sonomalavender.com

PHOTO: GARY ROSENBERG

"Lavender is one of the most useful and beneficial herbs known to mankind. Its soothing and relaxing qualities, antiseptic properties, and pleasing aroma make lavender highly desired worldwide."

– Gary Rosenberg, Sonoma Lavender

Stoechas

Angustifolia

Provence

Grosso

THE GENUS *Lavandula*

Lavender belongs to the genus *Lavandula*, part of the *Labiatae* family of mint, sage and thyme. Most lavender grown in America falls under four sections of the genus *Lavandula*: *Stoechas*, *Dentata*, *Latifolia* and *Angustifolia*.

Lavandula stoechas (*L. stoechas*), commonly called Spanish Lavender, was the most popular distilled oil in the Middle Ages, and is now prized for its long and recurring blooming season.

Lavandula dentata (*L. dentata*) is often called French Lavender. It is easily identified by its 'dentata' or toothed edged leaves.

Lavandula latifolia (*L. latifolia*) is the oldest known lavender, often called Spike Lavender. In ancient references it was called Spikenard or Nard.

Lavandula angustifolia (*L. angustifolia*) was nicknamed English Lavender because of its importance to the English lavender-oil industry, even though the plant is thought to have originated in the French Alps.

'Provence' and 'Grosso:' two of America's most common lavenders, are actually *Lavandins*, a sterile hybrid of *Lavandula angustifolia* and *Lavandula latifolia*. The botanical name of *Lavandin* is *Lavandula x intermedia*. The x stands for "cross." For example: *Lavandula x intermedia* 'Provence' or *Lavandula x intermedia* 'Grosso.'

GROWING LAVENDER

- Choose plants that have been proven to grow in your area.

- Choose a spot with full sun.

- The soil must drain well, because lavender does not like wet roots. Amend the soil as needed with sand or pearlite to increase drainage.

- The soil pH should be slightly alkaline, between 6.5 and 7.5. Amend the soil with lime, to add alkalinity.

- Plant new plants after any danger of frost.

- Trim flowers after the first year to send the plants energy to developing a stronger root system.

- Some drip irrigation is usually necessary.

- Overhead watering is not recommended.

- Prune back to the main plant in the Fall, if you have not have already done so when harvesting lavender stems.

- Fertilize with composted chicken manure and Bone meal when planting.

- Weed control is important to the beauty and health of lavender. Pull weeds within the plant. Between the rows can be handled by pulling, disking, weed wacking or applying weed cloth.

BONNY DOON FARM

SANTA CRUZ • CALIFORNIA

"Ye flowery banks o'bonie Doon,
How can ye blume sae fair;
How can ye chant, ye little birds,
And I sae fu' o' care!"

–excerpt from "The Banks O'Doon," by Robert Burns, Scotland, 1783

In 1972, the Moeller family divided their one hundred and fifty-two acre farmstead among their offspring. Frieda Moeller, an avid gardener, encouraged her daughter and son-in-law, Diane and Gary Meehan to start a lavender farm. Both of them had come from family backgrounds of farmers and gardeners.

In 1974, the Meehan's planted their first 2,000 lavender plants and "Bonny Doon Farm, America's Original Fine English Lavender Estate®" was born.

The Meehan's had great respect for the traditional paintings of lavender in England and Tasmania. They built their farm to traditional standards and planted every original plant and tree themselves. "It was a true labor of love on our fine agricultural land hidden within a beautiful forest that became the Bonny Doon State Nature Preserve," says Gary Meehan.

Currently, the farm has 4,000 plants in two varieties: *Lavandula angustifolia* and their own cultivar, *Lavandula 'Frieda,'* named after their inspiration, Frieda Moeller, Diane's mother. The Meehan's produce a full line of lavender products, including Gardener's Salve®, Eau de Cologne (lavender, of course, lavender shampoo and conditioner.

Not open to the public, Bonny Doon Farm is located northwest of Santa Cruz at an elevation of 1,550 feet. The redwood forested area known as Bonny Doon was founded in the 1850's as a logging area. John Burns, a Scotsman living there, named Bonny Doon after a line in Robert Burns' song 'The Banks O' Doon.'

"We still pinch ourselves at such good fortune," Diane says. "We deeply respect what was given us, and strive to provide a true quality product, without compromise."

http://www.bonnydoonfarm.com

LITTLE SKY LAVENDER FARM

BOULDER CREEK · CALIFORNIA

Maria and Rick Maze wanted to add culinary lavender to their pear and apple orchard, under a little piece of sky ringed with redwood, Douglas fir, live oak, tan oak, and madrone trees. They discovered that *Lavandula angustifolia* had a long history as a culinary delicacy, dating back most famously to Queen Victoria, who drank lavender tea daily.

L. angustifolia is one of the first true wild lavenders from which all lavenders have been cultivated (the others include *L. stoechas*, *L. dentata* and *L. latifolia*). Some purists insist on *L. angustifolia's* sweet heavenly aroma, preferring it to the stronger scent of *Lavandula x intermedia* hybrids (called *lavandin*) such as 'Provence' or 'Grosso'. The *L. angustifolia* scent is more delicate, and oil production takes four or five times the amount of the flowers as the more productive *lavandin* hybrids.

The Maze's bought *L. angustifolia* seed from four distinct seed stocks, including one from the Provence region of France, and planted them in egg crates. Over the next few months, they moved them to four-inch pots, then gallon pots, where they over-wintered. They tended the lavender starts for a year before putting them into the ground the following spring. In May of 2005 they planted 1200 carefully tended *L. angustifolia* plants.

What sets Little Sky Lavender apart is that their crop is grown from seed, distinctive from the cloned variety. With seeds, there are natural variations in the plants, unlike clones which are identical. The Mazes' goal is

PHOTO: DAVE MAZE

PHOTO: DAVE MAZE

PHOTO: LEIGH ANN MAZE-GESSNER

PHOTO: DAVE MAZE

"We planted seeds in seed trays, transferred to six-packs, and then to gallons, before planting 1,200 angustifolia into the ground."

–Maria Maze, Little Sky Lavender Farm

to provide the highest quality *L. angustifolia* culinary lavender for cooking. A historic staple in the households of Europe, and still used widely there, it is a unique experience in the culinary arts.

In addition to their unique *L. angustifolia* culinary lavender, the Maze's develop delicious mixes and kits including Lavender Cookie Mix, Lavender Brownie Kit, Culinary Lavender Sugar, Culinary Lavender Sea Salt, and Lavender Tisane, an herbal tea. They include recipes to use their unique herbs and spices. The delicious result is hand grown lavender blends that will enhance meals and recipes for chefs, novice to gourmet.

In 2010 Maria and Rick increased the number of plants and varieties and will be planting second generation seed starts into the ground in the fall. Dave, Rick's brother, has faced challenges with gophers, who have developed a love of the farm's tomato plants and flowers. This year more above ground container gardening will be the spotlight.

"We will conquer those rascals. We will prevail," Rick says.

Luckily, gophers don't like lavender and leave it alone.

http://www.littleskylavender.com

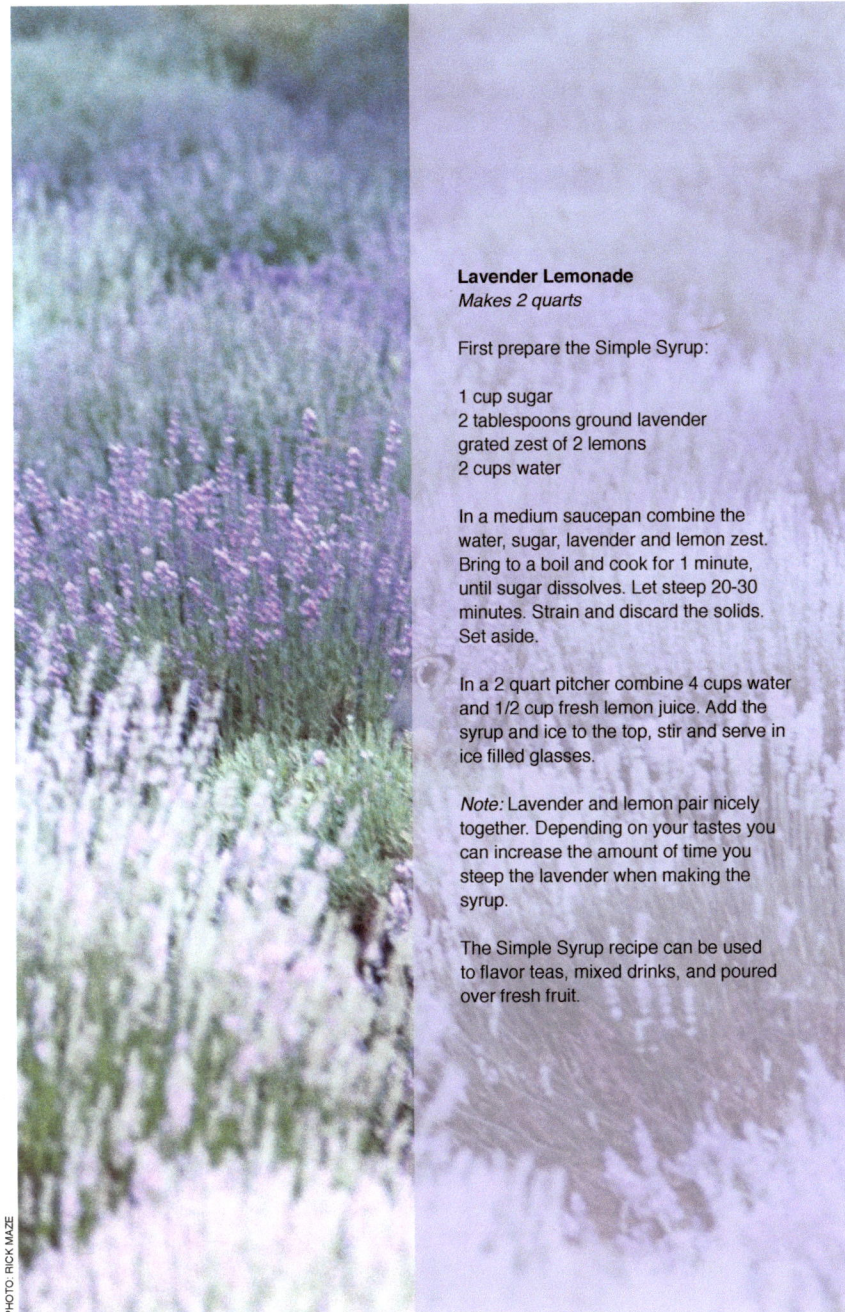

Lavender Lemonade
Makes 2 quarts

First prepare the Simple Syrup:

1 cup sugar
2 tablespoons ground lavender
grated zest of 2 lemons
2 cups water

In a medium saucepan combine the water, sugar, lavender and lemon zest. Bring to a boil and cook for 1 minute, until sugar dissolves. Let steep 20-30 minutes. Strain and discard the solids. Set aside.

In a 2 quart pitcher combine 4 cups water and 1/2 cup fresh lemon juice. Add the syrup and ice to the top, stir and serve in ice filled glasses.

Note: Lavender and lemon pair nicely together. Depending on your tastes you can increase the amount of time you steep the lavender when making the syrup.

The Simple Syrup recipe can be used to flavor teas, mixed drinks, and poured over fresh fruit.

CLAIRMONT FARMS

LOS OLIVOS • CALIFORNIA

Meryl Tanz awakened to hens clucking outside her French doors. She suspected Vera, a Spitzhauben German hen, had gone broody and had laid her eggs in some secret location. Meryl followed her and discovered the babies under a bush, two tiny black chicks with white spots on their heads.

When Meryl came back inside from her egg hunt, her husband, Glenn Thailheimer, brought her a steaming cup of coffee. They started their day's work with a discussion of repairs on the lavender's irrigation system which draws water from Lake Cachuma. They finished their coffee and headed out for morning chores.

In the year 2000, Meryl decided to leave the business of raising thoroughbred horses, but searched for something to do with her farm. Santa Barbara County buzzed with tourists visiting wineries, and Meryl thought they'd enjoy a different place to visit. She planted fields of lavender with a single cultivar, lavandin 'Grosso', prized for its high oil content and potent pine-y aroma.

Glenn made lavender essential oil for the night's internet orders of body lotion and goat's milk soap. He hauled down two boxes of dried lavender flower buds and filled the distiller with fifty pounds of buds. The 250-liter copper alembic, made in Portugal especially for distilling essential oils, steams the lavender buds and condenses the vapor, then separates the oil from the water.

People tease Glenn about whether he uses his still to make alcohol. "If I can earn $750 for a gallon of lavender oil versus $30 for a gallon of alcohol,"

PHOTO: GLENN THALHEIMER

Glenn asks, "which would you do?"

Even the distiller's castoff water is valuable. Called a hydrosol, it is infused with lavender oil and can be used for many things from a skin toner to linen spray.

Meryl and Glenn harvest some 'Grosso' by hand and hang it from the rafters to dry for crafts and bouquets. But much is harvested to distill for oil. They cut the lavender with gas-powered hedge clippers, harvesting and trimming the lavender bushes at the same time. Then the lavender is laid on a sheet and dried in the sun. Once it dries, they walk on it to separate the buds and rake off the stems. The buds are sifted to remove any leftover leaves and stems and then put into the copper still to steam.

"The best oil comes from lavender buds without stems because all the oil is premium," Glenn says. "It makes no difference whether we distill fresh flowers or dried. The proof is that lavender buds do a great job of retaining their scent twenty years later."

The 'Grosso' lavender oil is used in dozens of products but the one which Meryl is most proud of is the dog shampoo. "It repels fleas and ticks and soothes itchy skin," Meryl says. Clairmont Farms also provides a mattress company with loose lavender for aromatherapy and to repel bed bugs.

The driveway to Clairmont Farms is lined with olive trees planted by Spanish missionaries more than 180 years ago. Three hundred year old oak trees dot the grounds. The fields bloom from mid June to late July, which

varies with the weather. Hot weather brings an earlier bloom. Cool and foggy weather delays the bloom to late July.

Despite the hard work of farming and producing products and selling them, Glenn and Meryl feel blessed. "I have nothing to complain about," Meryl says. "Everyday I have people from all over the country stop by. They share their stories and feelings about America. I feel like I know them by the time they leave."

http://www.clairmontfarms.com

HOW TO CARE FOR YOUR 'GROSSO' PLANT

Lavandula x intermedia 'Grosso' is a sterile hybrid between *Lavandula angustifolia* and *Lavandula latifolia*, which was developed in Europe for its disease resistance and ability to produce more oil than other varieties. It does not produce seeds and can only be propagated through cuttings. The *Lavandula intermedia* hybrids have the frost hardiness of *L. angustifolia*, along with the camphory odor of *L. latifolia*. 'Grosso' is one of the tallest lavender varieties, making it useful in landscaping. It blooms once a year in June/July with deep purple flowers.

- Plant in sandy, well drained soil
- If in a pot, it will need to be a big one with holes in the bottom for drainage
- Watering once a week is usually enough
- More only if it is dry
- Water the soil not the top of the plant
- Full sun if possible, though some shade is okay.
- Soil should be slightly alkaline. Add lime if necessary.
- Cut the plant back to its core in the fall before it becomes dormant

KEYS CREEK LAVENDER FARM

VALLEY CENTER · CALIFORNIA

PHOTO: TRICIA METEER

PHOTO: TRICIA METEER

PHOTO: LAUREN LEMONS

"We're not in the middle of nowhere, we're at the end of nowhere," says Alicia Wolff, owner of Keys Creek Lavender in north San Diego County.

In search of a spiritual transition, Alicia Wolff and Chris Kurisu set out to find a retreat to expand their holistic lifestyle. They were drawn to Palomar Mountain, once the sacred home of the Temecula Indians. The Indians called this land Temeku, "the place where the sun breaks through and shines on the white mist." Alicia and Chris watched as the sun cleared the mist, and revealed fields of lavender. They had found their new home.

Alicia and Chris's vision for Keys Creek Lavender Farm is a place where people can feel connected with the earth. As visitors wander the 20,000 lavender plants on 8.5 acres, exploring over a dozen cultivars, they are engulfed with the enchanting scent. The aroma of the therapeutic herb works its relaxing, healing powers. Visitors always comment on how uplifting it is to be surrounded by lavender.

"I quickly realized that I needed to let go of old beliefs and patterns of how I thought business was done," Alicia says. "And get in sync with the rhythm and timing of Mother Nature."

Chris Kurisu conducts tours teaching visitors how lavender is cultivated, harvested, and finally distilled into pure essential oil. He also educates them about the many uses of lavender. Lavender arts and crafts workshops and cooking classes are also offered.

"Whether running my hands over the blooms, hearing the bees busy at work, or inhaling the sweet fragrance, I am in absolute awe of the beauty before me. I am humbled and inspired by the power of this herbaceous plant."

–Alicia Wolff, Keys Creek Lavender Farm

But perhaps unique to Key's Creek Lavender Farm, are the spiritually-oriented workshops which reflect Alicia's dedication to growth and spirituality. Seekers can attend "Flower of Life" and "Meditation and Chakra Clearing" workshops and related classes throughout the year.

"Walk in our living labyrinth or sit in our meditation garden," Alicia says. "Or plant a lavender plant in memory of a loved one in the peace garden. The color 'lavender' is for cancer awareness. A portion of our profits will be donated to cancer research."

In keeping with her mission of healing, sustainability, community and volunteerism, South African born Alicia Wolff dedicates Keys Creek Lavender resources to fundraising as diverse as horse rescue, to supporting children in South Africa with HIV/AIDS.

Keys Creek Lavender is open to the public in May and June when the lavender is in bloom. It has a gift store, tea house and distillery, where the essential oil is extracted from the plants. They make their own organic products which include body care, baby care, pet care, spa, home and culinary. One favorite is Sleep Balm, for those who have difficulty falling asleep and staying asleep.

http://www.keyscreeklavenderfarm.com

PHOTO: LEON METEER

PHOTO: TRICIA METEER

PHOTO: TRICIA METEER

LAVENDER ✴ LIFESTYLE

Lavender comes out of the closet.

Lavender freshens musty smells and repels moths and insects. Hang lavender sachets in your closet. Line your linen closet with lavender drawer liners. Layer lavender sachets with your sweaters and wool to avoid moth holes.

47

LAVENDER VALLEY

HOOD RIVER • OREGON

As Dayle Harris crested a hill on a drive through the Australian island of Tasmania, he saw a sight so spectacular he had to stop the car.

"Spread below me was a sea of brilliant purple as far as the eye could see." Dayle stared at the endless field for half-an-hour until the desire to know what it was overwhelmed him. He drove into the nearest township and stopped in a cafe. The locals laughed at his amazement and told him the remarkable fields were planted with lavender.

Dayle, a commercial pilot and his wife, KaiLai, a flight attendant, flew to many countries around the world for their jobs. The couple made a point of visiting other lavender farms and shops, many in England. They gathered growing tips and ideas for crafts, culinary uses, and products with the dream of starting a lavender farm when they retired from flying.

They settled in Oregon, with its pure glacial streams, abundant wildlife and a history of successful farming. Their twenty one acre Lavender Valley Farm is located in the rich and fertile Hood River Valley between two active volcanoes; Mt. Hood and Mt. Adams. Because of the unique micro-climate the winter months are not extremely cold and the summer months are dry and warm, making an ideal climate for lavender. The latitude of forty three degrees north is approximately the same as that of southern France where lavender is grown commercially, making the climate excellent for lavender production.

In 2001 they planted lavender and pear trees. Now they have 1,800 Bartlett and Anjou pear trees and more than 10,000 lavender plants. The

PHOTO: JUDY GALLOWAY

"Spread below me was a sea of brilliant purple as far as the eye could see."

–Dayle Harris, Lavender Valley

first two seasons they pruned the blossoms from the lavender plants to ensure a strong root structure. They harvested their first commercial lavender crop in the summer of 2004. The farm grows *L. angustifolia* 'Royal Velvet' and 'De Lavande' and *lavandin* 'Provence' and 'Grosso.'

They distill oils from both *angustifolia* and *lavandin*. The fragrance of lavender is contained in its oil, which the plant produces and stores in tiny cells at the base of each floret. The lavender essential oil is separated from the flowers by steam distillation. The distillation kettles are packed full and steam is passed through the plants for about forty-five minutes. The mixture of steam and oil vapor is piped to a condenser, where the pure essential oil collects in the separator. Various lavender cultivars produce differing amounts of oil. The spent lavender is cooled and used as mulch. The distillation unit operates daily at the farm, from the beginning of harvest in July until mid-September.

The amount and quality of the oil depends on the type of lavender cultivar and also on the amount of sunshine just before harvest. The hybrid *lavandin* produces 4-5 times more oil than *angustifolia*, making *angustifolia* essential oil more costly. *Lavandin* oil contains more camphor than *angustifolia*.

Visitors to Lavender Valley Farm can take a guided tour in the summer months from the Harris's son, Jonathan. They can visit KaiLai and learn about her lavender lotions, soap, creams, linen sprays, sachets, dried flowers and honey. Dayle's ninety two year old mother, nick-named Bubba,

hand paints pictures and glassware. The farm sells more than sixty varieties of lavender plants to the public, and they ship within the United States.

If one is lucky enough to visit the farm during the Lavender Festival in the Gorge, they might hear Jonathan playing 'Purple Rain' on their Stroud antique piano.

http://www.lavendervalley.com

MAGIC OF LAVENDER

Lavender oil improves the feeling of well being.

Ever hear of dopamine? Dopamine is a chemical compound released in the brain in response to positive experiences such as food, sex, friendship, drugs, which gives one a feeling of reward and well being. In a study by the Korean Department of Health and Keimyung University, Lavender has proven to affect dopamine release, creating that positive feeling.
*Department of Public Health, Keimyung University, Taegu 704-701, Republic of Korea, 26 June 2009.

Change a cranky mood by adding lavender to your day. Start off by showering with Lavender Shower Gel and following it with Lavender Body Lotion. Line your drawers and closet with lavender to infuse clothes with a feel-good scent. At the end of a stressful day wrap up with a lavender neck pillow or heat wrap.

IN THE LAVENDER HAZE

You are in love if you are "In the Lavender Haze." The phrase was coined in the fifties, but love and lavender have been bedfellows long before that. Lavender has legendary romantic powers.

Cleopatra is said to have seduced Julius Caesar and Mark Antony wearing lavender perfume. In Tudor times, damsels drank lavender tisane—lavender flavored tea—at bedtime, and called on St. Luke to reveal the identity of their true love:

'St. Luke, St. Luke, be kind to me.

In my dreams, let me my true love see.'

The Bible mentions lavender many times, as spikenard, meaning 'pure'. Mary used lavender oil to anoint the feet of Jesus as a supreme act of motherly love. Judith used it to seduce Holofernes. The Queen of Sheba offered lavender to King Solomon.

Lavender is called the 'flower of love' or the 'flower of distrust', depending on who you believe. ('Flower of distrust' stems from the fact that lavender's fresh scent masks foul odors.) Victorian girls tucked lavender sachets between their breasts to lure suitors. Pillows filled with lavender flowers promoted romance, and lavender kept under the beds prolonged the ardors of marriage. This European tradition was proven to have a base in reality when the Chicago Institute of Taste and Smell discovered the scent of lavender and pumpkin arousing for males.

PHOTO: REBECCA GOSSELIN

Lavender has a wicked reputation for attracting lovers and spurring passion. Bathe in a lavender bath before a love tryst. Scent the bed with lavender spray mist. Massage a lover with lavender oil. And don't forget to sprinkle lavender water onto your lover's head to keep him faithful.

Lavender is often the flower of choice at weddings. Brides weave lavender into white roses for their bouquets. Ring bearers carry wedding rings on lavender pillows. Guests throw lavender instead of rice to wish the couple good luck and a long marriage.

'In the Lavender Haze.' Doesn't it sound like a lovely place to be?

OREGON LAVENDER FARM

OREGON CITY • OREGON

Jim Dierking's story is a serendipitous case of the tail wagging the dog. Jim, long enamored with essential oils and aromatherapy, owns Liberty Natural Products, distributing more than 1,000 botanical extracts around the world. But in 1999, he needed a bigger place for his growing business and bought a rambling ninety-acre farm which had been a Carnation Egg Ranch and later a Rex Rabbit operation. It had fallen into disrepair. With 150,000 feet of barns and buildings, the farm offered him a chance to renovate the property into a fine facility for his business and plant fields and fields of lavender.

"I chose to plant lavender because women love it," Jim says, smiling. "Besides the fact that it is the most highly regarded and commercially grown oil in the world."

Several friends and colleagues have guided Jim through the ambitious venture.

Dr. Don Roberts, a retired Oregon State University professor, had developed a highly regarded *Lavandula angustifolia* variety, 'Buena Vista', prized for its vibrant blue color and high quality essential oil. Don helped Jim select the varieties of lavender to plant and educated him on cultivation.

Jim was also mentored by Robert Unrath, a flavor and fragrance chemist, and head of the U.S. distribution arm of the French company,

"The healing and aroma of lavender led me into manufacturing and distributing essential oils around the world."

–Jim Dierking, Oregon Lavender Farm

LAVENDER ❋ LIFESTYLE

Travel with Lavender.

Lavender reduces stress and provides practical solutions for your travels. Lavender-filled slippers soothe tired feet. Lavender Travel pillows and sleep masks relax and rejuvenate. Lavender stuffed animals soothe cranky kids. Lavender spray mist refreshes psyche and hydrates skin. Lavender essential oil heals cuts, relieves pain, relaxes, aides sleep, bans odors.

whose family's great grandfather founded the lavender industry in France. Robert provided Jim with an overview of large-scale lavender cultivation and distillation, and arranged a trip to Provence. The trip inspired Jim to develop the agri-tourism aspect of the farm.

Oregon Lavender Farm's manager, Ivan Grier, has been with Jim since they planted in 2002. Straight out of the horticulture program of Mt. Hood Community College, Ivan helped Jim clear the brush, pull stumps, and remove rocks. Eventually, he learned propagation, cultivation, and distillation of the lavender oil.

Today, Oregon Lavender Farm has 60,000 plants on twenty acres, and Jim plans to expand. They grow *Lavandula angustifolia* 'Buena Vista', *Lavandula x intermedia* 'Grosso', and *Lavandula latifolia*. They have a small scale distillery and are developing a larger distillery to handle their expanding lavender fields.

"My vision is to become the 'Mother' distillery for the Oregon lavender industry," Jim says.

But Oregon Lavender Farm is not all business, proven by the water and light show at the base of the aromatherapy garden. The water cascades down a hillside to a large pond behind a stage and dance floor where fifty water jets and two hundred lights are synchronized to music. In late June, Oregon Lavender Farm hosts the Clackamas County Lavender Festival. Visitors enjoy lavender honey chicken wings, lavender beer, and live music while picking lavender and learning about sustainable agriculture and oil distillation.

http://www.libertynatural.com

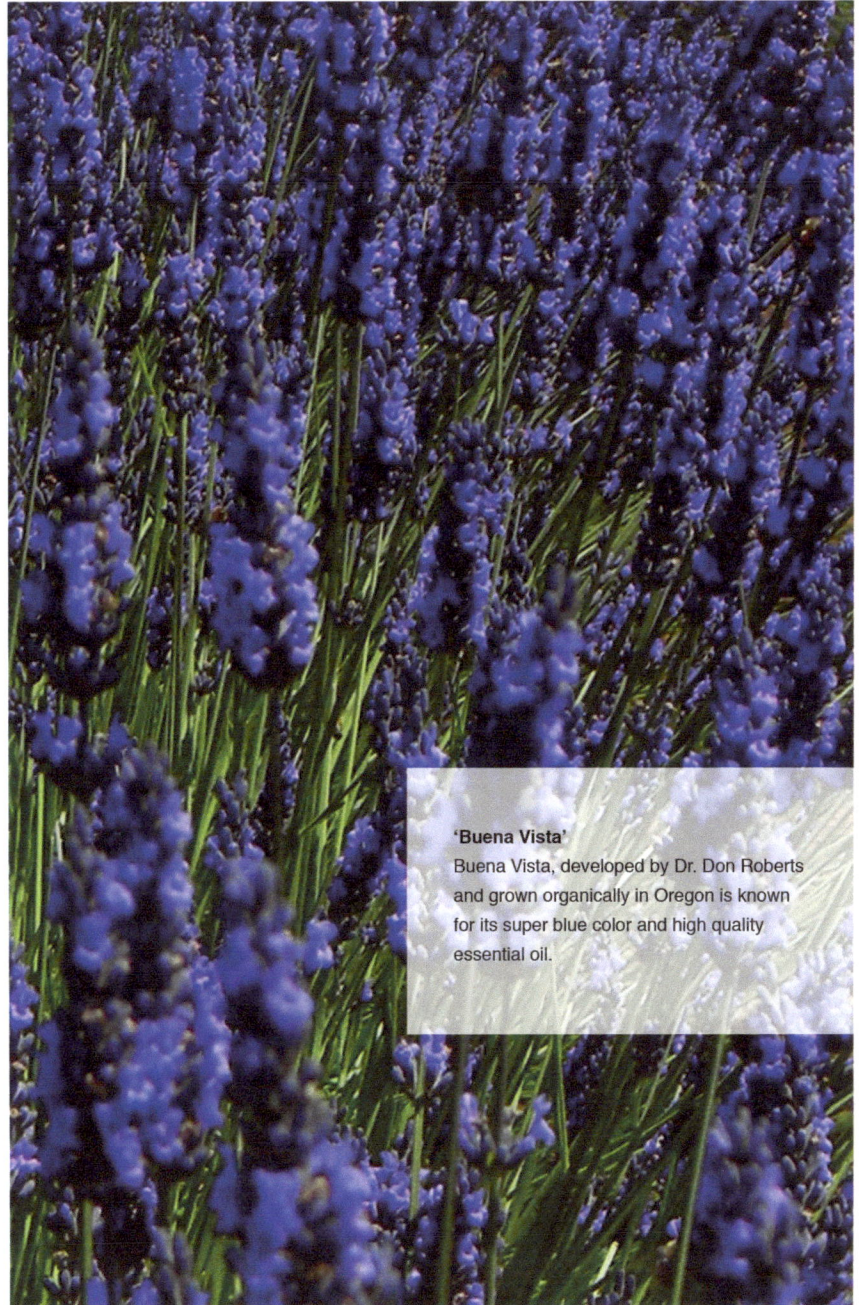

'Buena Vista'
Buena Vista, developed by Dr. Don Roberts and grown organically in Oregon is known for its super blue color and high quality essential oil.

PELINDABA LAVENDER FARM

SAN JUAN ISLAND · WASHINGTON

"Lavender has a two-thousand year old history as having more uses than any other plant."

–Stephen Robins, Pelindaba Lavender Farm

Stephen Robins has come full circle. As a child growing up in South Africa, Stephen enjoyed vacations in the solitude of the bushveld. Far from that peaceful life, as a writer, editor, and practicing physician, Stephen later began his own medical communications organization, catalyzing the establishment of many national and international healthcare standards. He dreamed of eventually retiring to a place of natural beauty with nourishing physical and spiritual benefits, and bought twenty acres on San Juan Island in Washington, with stunning views of lakes and the Olympic Mountains. By his mid-fifties Stephen moved to San Juan Island full-time and contemplated what to do with the property. It seemed natural for him to settle in this peaceful place. He named it 'Pelindaba', Zulu for 'Place of Great Gatherings.'

Stephen wanted to make Pelindaba a place where people could soak up nature. At the same time, he wanted it to be self-sustaining. The search for a suitable crop led him to lavender. Lavender could survive without irrigation in the dry climate with less than twenty inches annual rainfall. No artificial fertilizers would be needed in the rich soil of an ancient river valley. No one was growing lavender on the island, allowing him not to compete with someone farming for a living.

Stephen planted 2500 plants in 1999. His original vision was to farm and sell his lavender but economic evaluation led him to build a large-scale distillery (500 gallon capacity), distill large quantities of essential oil, and

PHOTO: DANA STYBER

PHOTO: DANA STYBER

PHOTO © PELINDABA

PHOTO: DANA STYBER

PHOTO: MARY IRELAND

develop products using oil and the flower buds. As the years progressed, he needed more lavender. Today, Pelindaba has 25,000 plants on sixteen acres, including *L. x intermedia* 'Grosso,' *L. x intermedia* 'Provence,' *L. angustifolia* 'Royal Purple,' *L. angustifolia* 'Hidcote,' and *L. angustifolia* 'Munstead.' The property includes a farm store, educational visitors' center, distillery, drying barn, production center, certified kitchen, and residences for the farm manager and owner. Pelindaba Lavender crafts more than 250 lavender products from its organic lavender: decorative, personal care, therapeutic, culinary, household care and pet care.

The farm is open to the public April through October. Operations have expanded to a Friday Harbor store and licensee stores off the island.

For Stephen, the farm is a constant source of reward. "At Pelindaba Lavender, we make all our products solely from the certified organic lavender we grow ourselves—and we do all this right on the farm."

Stephen embraces the fact that retiring, at least for him, is not a period of relaxation. Since he started Pelindaba Lavender Farm, he has learned about farming, distillation, manufacturing, retailing, wholesaling, and agri-tourism. "In short," he says, "retirement reinvented."

http://www.pelindabalavender.com

PURPLE HAZE LAVENDER FARM

SEQUIM • WASHINGTON

Washington State Park Ranger Mike Reichner bought a 2.5 acre cow pasture in Sequim, Washington, to build a home. Mike's farm boy roots gave him a strong desire to grow something, although he had no idea what it could be. He attended a meeting about growing herbs commercially and was struck with the idea of growing lavender. After researching lavender all winter, he planted a few dozen lavender varieties which did well. He planted a few hundred more in the fall.

In 1997, Mike opened the farm to the public to pick their own lavender bundles. Next, Mike had a website built for Purple Haze Lavender. "Now the whole world could visit Purple Haze Lavender Farm," Mike said with a smile.

That fall a writer from *Forbes* Magazine, Anne Linsmayer, wrote a small article which included a photo of Mike and his wife. "Suddenly, our world changed. The day after Forbes hit the newsstand, I was on Paul Harvey News. That was just the beginning." Mike found himself in the national spotlight, which forced a tough decision to leave his park ranger job of twenty years, forfeit his retirement and become a full time lavender farmer and media darling.

Mike spoke about agri-tourism and became a consultant for dozens of aspiring lavender farmers across the country. Purple Haze has been featured on many national television programs and appeared in national and regional magazines.

Now approximately 150,000 people visit Purple Haze during the bloom season, April through Labor Day. Mike has expanded to 30,000

PHOTO: LIGHT RAIN PHOTOGRAPHY

"I recall very clearly all of my friends and family laughing at me over this idea. I kept plugging along."

–Mike Reichner, Purple Haze Lavender

PHOTO: LIGHT RAIN PHOTOGRAPHY

lavender plants on 12.5 acres, including 'Grosso' and 'Hidcote Giant' *lavandin*, and 'Royal Velvet,' 'Melissa,' and 'Sachet' *angustifolia*. He and his crew harvest thousands of bouquets of fresh and dried lavender and also distill their own lavender essential oil. They are on their third still, a propane-fired, 1.2 million BTU unit, which they use to distill oil for themselves and neighboring lavender farms. "It literally makes the ground tremble when it's operating," says Rosalind, Mike's wife.

Purple Haze makes about a hundred lavender products, featuring a culinary line of five flavors of ice cream, a lavender coffee, lavender chocolate bars, herbs de provence, and lavender hot chocolate mix. Rosalind says their lavender dressing is the most popular. They also offer soaps, lotions, and snuggle wraps.

http://www.purplehazelavender.com

LAVENDER ✲ LIFESTYLE

Start your day with Lavender in the bathroom.

Lavender is antiseptic, anti-inflammatory and heals dry, chapped skin. Lavender soap kills bacteria, smoothes the skin, promotes cell growth. Lavender Shower Gel or Bubblebath rejuvenates you and elevates your mood, while cleansing and soothing the skin. Lavender Body Lotion soothes and moisturizes. Lavender Hand Crème and Footbalm softens super rough skin in tough places. Give a massage with lavender massage oil.

PHOTO: GARDENPHOTOWORLD

Reduce feelings of conflict and anxiety with lavender.

Sometimes things just don't seem to go your way, leaving you with angry, anxious feelings. Now a study from the University of Tsukuba in Japan concludes that lavender can significantly change those feelings of conflict and anxiety.

Lavender oil produced significant anti-conflict effects in the Geller and Vogel conflict tests, suggesting that the oil has an anti-anxiety effect. Linalool, a major constituent of lavender oil, was identified as the pharmacologically-active constituent involved in the anti-anxiety effect of lavender oil.

National Institute for Environmental Studies, 16-2 Onogawa, Tsukuba, Ibaraki 305-0053, Japan College of Nursing and Medical Technology, University of Tsukuba, Japan 14 December 2006. SCIENCEDIRECT.COM

Reduce feelings of conflict and anxiety with a lavender candle or lavender diffuser. Tuck a lavender sachet in the car or office. Spray lavender whenever conflict arises.

PHOTO: GARDENPHOTOWORLD

PHOTO: GARDENPHOTOWORLD

BECKER VINEYARDS' LAVENDER FIELDS

STONEWALL • TEXAS

"Bring the things you love around you and let nothing get in your way."

– Dr. Richard Becker, Becker Vineyards

The complex sweetness of Viognier grapes led endocrinologist and winery owner Dr. Richard Becker and his wife Bunny to be the first to plant Viognier in Texas. They visited France to study about the ancient varietal. The southern Rhone River Valley of Provence with its hot, dry summers and sandy soil reminded the Beckers of their vineyard in Hill Country, Texas with one stunning difference. Rows upon rows of vibrant lavender stretched from the country roads, dazzling their eyes and filling their noses with pungent perfume.

The Beckers were hooked. They returned to their vineyards and with help from their daughter Clementine planted 10,000 lavender plants in 1998, including *lavandins* 'Grosso,' and 'Provence,' *L. angustifolia*, and *L. stoechas*. They embellished their vision with the artistry of Impressionist painters, adding yellow sunflowers, red poppies, blue bonnets, and artichokes. The three acre lavender field ran alongside the Becker's winery. The 10,000 square foot reproduction of a nineteenth century German limestone barn was surrounded by peach tree orchards, a milk house, a well, a windmill and an 1890's log cabin.

But Texas weather visited a nightmare on the Becker's lavender vision. In 2000, they planted 700 new plants and lost 85 percent of them to a heat wave and drought. In 2002, thirty inches of rain fell in a week and killed 60 percent of the plants. The Beckers kept searching for varieties that could withstand the wild weather. They determined that 'Grosso' was not a good variety for

Texas, but Stoechas (Spanish) did much better.

"One year we invited eighty people for a lavender luncheon on the porch," Bunny Becker says. "We were all looking out at the blooming lavender fields when a massive cloud of grasshoppers descended on it. We heard the screech of their wings and a snip, snip, snip. Lavender heads drooped uneaten but clipped at the base. They tried to eat the lavender but apparently didn't like it. The guests witnessed the destruction sadly, but were more horrified when the grasshoppers moved onto the porch. My husband Richard raised his glass in a toast. 'I would like to welcome everyone to our first-ever Grasshopper Luncheon.'"

Now visitors to Becker Vineyards enjoy the rewards of the Becker's unwavering determination. They taste award winning wines from the winery and gaze on their lavender fields, trimmed to a more manageable 4000 plants. During the lavender bloom, April through June, people wander through the rows to pick a bunch while accompanied by live music, impromptu singing, and dancing. Cheese, charcuterie, and lavender-inspired cuisine are available to savor with the wine. Chris Perrenoud, the lavender manager, offers a myriad of lavender products in the Becker's Lavender Shop, including handmade lavender soap, linen spray, lotion, eye pillows, sachets, and candles.

The Becker Vineyards Lavender Fest usually takes place the first weekend of May. The event features lavender educational speakers and vendors promoting lavender and other herb-related products. Gardening tips, cooking demonstrations, wine tasting, and luncheons are also part of the festivities.

http://www.beckervineyards.com

PHOTO: CHRIS PERRENOUD

HILL COUNTRY LAVENDER

BLANCO • TEXAS

> *"The weather is a major challenge in Texas. One year, we face extreme drought, the next year, we are flooded. But once in a while, we get the perfect season where the lavender is beyond beautiful."*

–Tasha Breiger, Hill Country Lavender

In 1999 high school junior Tasha Brieger became a photo assistant to Robb Kendrick, an award-winning photographer for *National Geographic*. Robb had photographed the lavender fields of Provence for a book about perfume making. He noticed the similarities between the climate of Provence and his home in central Texas and wanted to recreate the beautiful hills he'd captured for the book. Robb and his wife, Jeannie Ralston, a New York writer and editor, started Hill Country Lavender, the first commercial lavender farm in Texas. (Jeannie Ralston wrote a memoir about starting Hill Country Lavender, *The Unlikely Lavender Queen: A Memoir of Unexpected Blossoming*.)

After working for Robb a short time, Jeannie asked Tasha to help with the lavender farm. Tasha became immersed in the development and expansion of the lavender product line, and became the farm's manager in 2004. She learned all aspects of lavender, from cultivating, to harvesting, to making products.

But nothing could have prepared Tasha for the events at the end of her junior year in college. While working to finish a double major in graphic design and photography, Robb and Jeannie offered Tasha the chance to purchase the business. Having worked in all aspects of the operation for six years, and with support from her entrepreneurial parents, Tasha purchased Hill Country Lavender in 2005.

Hill Country Lavender is a two and a half acre farm with a panoramic

view of undulating land and beautiful live oaks. Their 2,500 plants include 'Provence', 'Sachet', 'Munstead' and 'Stoechas' cultivars. Visitors are invited to cut their own lavender throughout the lavender season, from mid-May until July, with peak color from early to mid June. The farm store offers seventy lavender products, including soap, sachets, oil, linen spray, bubble bath, and eye pillows.

"Lavender is such a wonderful plant and so very versatile," Tasha says. "You can make wonderful body products, or use it in savory or sweet recipes, or you can enjoy the beauty of the plant in full bloom. What could be better than that?"

Hill Country Lavender participates in The Blanco Lavender Festival in June, featuring tours of local lavender farms, a lavender and craft market, and a speaker's pavilion.

http://www.hillcountrylavender.com

Lavender aromatherapy has relaxation effects that lower cortisol and has beneficial effects on coronary circulation.

Mental stress is an independent risk factor for cardiovascular events and impairs coronary circulation. Lavender aromatherapy is recognized as a beneficial mental relaxation therapy. However, no previous study has examined the effect of this therapy on coronary circulation. A recent study assessed the effect of lavender aromatherapy on coronary circulation by measuring coronary flow velocity reserve (CFVR) on young healthy men, range 24–40 years. Simultaneously, serum cortisol was measured as a marker of stress hormones. The same measurements were repeated in the same volunteers without aromatherapy as a control study.

Serum cortisol significantly decreased after lavender aromatherapy, but remained unchanged in controls. In addition, CFVR significantly increased after lavender aromatherapy, but not in controls. Conclusion: Lavender aromatherapy reduced serum cortisol and improved CFVR in healthy men.

aDepartment of Cardiovascular Science and Medicine, Chiba University Graduate School of Medicine, Japan, Green Flask Laboratory, Tokyo 152-0035, Japan, Faculty of Pharmaceutical Sciences, Toho University, Japan 8 August 2007.

Treat your heart to relaxing lavender aromatherapy. Keep a lavender sachet ready to squeeze for a fresh dose of lavender aromatherapy. Use a lavender eye pillow for a short break in a hectic day. Burn pure lavender essential oil candles.

LAVANDE

BELLVILLE · TEXAS

"I planted the lavender in 'S' curves, mirroring a woman's body."

–Craig Stewart, Lavande

Opposite: Lavande lavender field in November, before winter pruning.

"If you tell a Texan he can't do something, he'll come back with 'I'll show you,'" explains Dana Stewart. "And, Craig is Texan through and through."

Texas is famous for longhorn cattle, oil, wide-open spaces, trucks, boots, and country music. The rolling hills of the Stewarts' sixty-five acre ranch were meant for running cattle and growing hay and sprouting a carpet of blue bonnets in the spring. But when Craig returned from a photography trip to Provence for his stationery line, he imagined lush purple spikes rustling in the breeze against a turquoise Texan sky.

He discovered a worthy opponent in his quest. The land and climate were hostile to lavender. The soil was clay, while lavender needs a porous soil. The pH was acid, but lavender prefers alkaline. The humidity was high, and lavender prefers a dryer climate. Craig Stewart, an artist and photographer, sculpted his new vision with one hundred tons of sand for drainage, one hundred tons of crushed limestone to add alkalinity, and miles of irrigation lines to transform their cattle ranch in Austin County into a lavender farm.

"In kindergarten Craig couldn't color within the lines, wanting to draw his own creation," Dana says. "But he loved show and tell."

Craig's 3,500 lavender plants don't follow traditional straight rows. He formed rows with 'S' curves, mirroring the form of a woman's body. "Not only

is it artistic, but good for land conservation," says Craig with a smile.

He became the farmer, producer and marketer, with the help of his father, Ed Stewart, Lavande's ranch foreman and overseer. Dana is his cheerleader, teaching full-time in Houston. Their daughters, Zoe and Lauren, help with marketing and special events.

Craig loves to show off his Texas land, a colorful canvas of painted sky and endless hills stretching from horizon to horizon. He is always ready to come in from the fields to explain farming to visitors.

Initially he planted lavandins 'Provence' and 'Grosso,' but 'Grosso' did not survive. The culprit was phytophera, a fungus that has become their enemy. They immersed themselves in a crash course in chemistry, choosing appropriate fertilizers and fungicides to fight the foe. The techniques that worked in other parts of the country proved ineffective for them. They now grow 'Provence,' 'Goodwin Creek Grey' and 'Royal Velvet.'

Craig says, "Guests come to see the lavender fields but become mesmerized with the different benefits of lavender." One favorite in the summer is Lavande's lavender mosquito spray. "In Texas we call mosquitoes Texas songbirds; they are so big. People love our lavender repellant."

The Stewarts built Lavande Pavillion for events in the style of a traditional Texas 'Ag Hall', while adding charm and comfort to their venue. They offer relaxed hospitality to each guest that graces their doorway. Visitors sit in rocking chairs on the wraparound porch, enjoying the constant breeze. Craig and Dana Stewart see the contented look in their guest's eyes and enjoy their true work of art.

http://www.lavandetexas.com

LAVENDER * LIFESTYLE

Lavender helps you relax and aides sleep.

Decorate beds with lavender-filled pillows to relax you and help you sleep at night. Or tuck a sachet in your pillow. A study by the Institute of Taste and Smell determined that men are aroused by the scent of lavender. Spritz your sheets with lavender spray mist.

RED ROCK LAVENDER

CONCHO · ARIZONA

*"Growing lavender inspires and challenges us.
It is our passion, livelihood, and a connection to meeting wonderful people."*

–Mike and Christine Teeple, Red Rock Lavender

"It's all in your altitude," Mike Teeple says.

Red Rock Ranch and Farms, owned by Mike and Christine Teeple, is at a lofty elevation of 6,100 feet in the White Mountains of Arizona. Mike points out the similarity to lavender grown in the mountains of Provence, on the Plateau de Vaucluse on Mt. Ventoux, at an altitude of 6,273 feet.

Lavender grown at these altitudes contains a higher ester content than lavender grown at lower altitudes. The environment and ecosystem of the White Mountains is optimal for producing intensely fragrant lavender with a scent that endures.

The Teeples initially planted lavender to landscape their 130 acre ranch because rabbits and deer don't like it. And while lavender deters insects one might consider bad, it also attracts beneficial bugs, butterflies, bees, and hummingbirds.

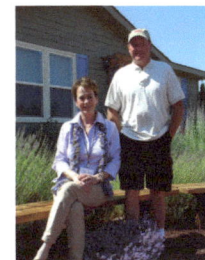

Now Red Rock Lavender Farm has more than 35,000 lavender plants in twelve different varieties. Their *lavandins* include 'Provence' and 'Grosso.' Their *Lavandula angustifolia* includes 'Royal Velvet,' 'Buena Vista,' 'Vera,' and 'Pink Melissa.' In addition, Red Rock Ranch and Farms propagates 10,000 to 15,000 plants per year in its greenhouse for people who want to start their own lavender farms and southwest gardeners who need plants acclimated to similar climates and altitudes.

Lavender not only loves the altitude of the White Mountains but also

PHOTO: CHRISTINE

the climate. The plants thrive in virgin soil, clean air, and pure spring water. Red Rock's lavender plants are irrigated by a highly economical automated drip system with each plant requiring about one gallon of water per week.

"At the time we started it, people hadn't been planting in higher elevations. As it turns out, lavender is very tolerant at this elevation," says Mike. "We let them go through some harsh winters and summers. They did well because lavender likes to go dormant," Mike says. "When dormant, their strength is not put into stems and leaves, but into strengthening the root system. This makes them hardier plants. Lavender plants grown at high altitude outlive lavender grown at lower altitude."

"High altitude lavender produces three to four hundred spikes per plant which is unusual because most plants yield about two hundred spikes," Mike says. "High altitude lavender holds its flowers better which allows for a later harvest. In coastal areas, they have to harvest them when they are about three quarters budded. If they harvest later when the flowers are coming out the flowers fall off when it's dried. The health and strength of our plants allows us to almost let the whole flower come out before we harvest."

At harvest time, they use two methods for drying. One method is to cut and dry the lavender in the fields. This lavender will be used to make the lavender oil. Other varieties, prized for their dark flowers, such as 'Royal Velvet' and 'Grosso,' are cut in bunches and hung upside down to dry in the barn. The dark barn helps keep the deep purple color.

The Teeples produce 100 percent pure high altitude lavender essential oil which some consider exceptional, and rivals Provence's high altitude 'Super' lavender.

"High altitude lavender essential oil has a well balanced fragrance superior to many others and is often the choice of aromatherapists," Christine says.

They create many products from their oil and dried lavender, including culinary/herb blends for cooking, spray mists, and their signature lavender shea butter cream.

http://www.redrocklavender.com

PHOTO: CHRISTINE TEEPLE

CAROUSEL FARM LAVENDER

MECHANICSVILLE • PENNSYLVANIA

"It's gratifying to see the plants turn green and produce something you can use or eat. It reminds us that plants are also living, breathing things."

–Niko Christou, Carousel Farm Lavender

A baby donkey had been born the night before, and Niko Christou tended it and its mother in the 18th Century stone barn. Black and white llamas and Scottish Highland cattle grazed in the pasture. Chickens strutted and pecked next to the barn and fieldstone farmhouse. A lavender-scented breeze wafted across the 35-acre grounds, the bucolic setting at once serene and inspiring.

Niko Christou, Manhattan photographer, and David Braff, Wall Street attorney, purchased the historic Buck's County property in 2000 as a weekend retreat, to ride horses and to get away from the noise of the city. Established in 1748, it had once been a dairy farm and later a horse farm. Then, in mid-20th Century, it became Carousel Farm, named for the exotic stage animals kept there for use on Broadway and at the Met.

Niko and David wanted to restore the farm's rural beauty, yet enhance it with something more. Inspiration struck when they traveled the fragrant Provence countryside, where centuries-old fieldstone barns and farmhouses graced a landscape of old grape vines and lavender fields. Carousel Farm, with its fieldstone farmhouse, 18th Century stone barn, and rolling fields, seemed the perfect place to replicate the south of France.

What started as weekend getaway became a labor of love and dedication for Niko. He had grown up cultivating olive trees and pressing oil on the Mediterranean island of Cyprus and felt his experience would help him farm lavender. But the harsh Northeast climate proved a formidable

challenge. He tested several varieties over the first few years and lost thousands of plants to freezing temperatures. After much trial and error, he settled on four varieties which thrive in Pennsylvanian soil: two lavandins, 'Provence' and 'Grosso' and two *L. angustifolia*, 'Munstead' and 'Hidcote,' from which he gets a second bloom.

"Different varietals flower at different times, so by planting four, I have color from June through October," Niko says. Peak bloom is usually the third week of June.

"I learned the hard way that snow is a friend to lavender," Niko says. "The weatherman predicted twenty-four inches of snow, so I set out to cover the plants with garbage bags to protect them, but only finished half the field before the blizzard hit. In the end, there was no difference between the lavender covered by bags and lavender exposed to the snow. The snow actually insulates the plants from wind and temperature."

Niko feels it is important to buy plants that have been propagated from plants acclimated to the region's particular weather. The same varietals imported from elsewhere will not be acclimated.

Niko Christou's Tips on Lavender Planting

A lavender plant needs two things: plenty of sun and dry conditions. It needs at least five hours of direct sunlight a day and should be planted in an area where the soil drains well.

It's important to buy varieties acclimated to your specific climate. Niko recommends planting *Munstead, Hidcote, Grosso* and *Provence* for the Northeast. "Before the first frost, cut any new growth back to where it started in the spring," Niko says. "If you don't, the plants can be split apart by the snow."

Carousel Farm uses *L. angustifolia* lavender for drying and cooking. They harvest lavandins primarily for oil because they yield four times the amount. During harvest, their thirty gallon steam distiller works constantly. They distill the lavender in small batches and monitor it closely. Niko feeds the distiller with freshly harvested lavender flowers, leaves and stems, yielding about 100 gallons of oil a year. Carousel Farm is known for its Lavender Essential Oil and Lavender Hydrosol. Niko uses the lavender essential oil in the farm's line of products, which includes soaps, lotions, and candles.

For Niko, it's not about the bottom line or growing his business. Whether he is feeding his newborn donkey, tending the lavender fields or capturing the fragrant landscape on film, Niko wants to share the tranquility of his farm with the visitors who come to experience it.

http://www.carouselfarmlavender.com

LAVENDER ✳ LIFESTYLE

Embrace a calm and relaxing way of living life.

- Infuse lavender into your environment. Lavender sprays, diffusers and candles create a tranquil, easy going atmosphere.
- When stress threatens, rub lavender oil on your temples, or squeeze a lavender sachet and inhale its calming scent.
- Unwind with a heated lavender neck pillow or spa mask. Lavender's potent aromatherapy soothes jangled nerves and elevates your mood.

LAVENDER BY THE BAY

EAST MARION • NEW YORK

Serge Rozenbaum is a French transplant. Born and raised in Paris, his first encounter with lavender was from a street vendor in Montmartre, the neighborhood he grew up in. The man led a donkey saddled with lavender bunches on the sides. His mother used to send him out to buy lavender, which she used to freshen the linens and flavor her cooking.

"In France, everyone grows lavender," says Susan Rozenbaum, his wife. "It is good for the body and the mind."

The Rozenbaum's have grown lavender for twenty years. First, in their yard in Southold, New York and now on the seventeen acre farm.

"When guests came on the weekends, we would hand them a pair of scissors to cut lavender and they would take home as much as they wanted," Susan said. "We realized we had more than we could ever give away, so we put a little stand at the end of the driveway with a sign 'Lavender for Sale.' We went out to the beach and when we came back, there was money in the can." A light went on.

Serge and Susan planted 50,000 lavender plants in twenty varieties. "We learned a lot about what to do and what not to do," Serge says. "The soil can't be compacted with clay. Good drainage is essential. And if the soil is too acid the plants will die."

He tells a story about heavy rains raising the water table and killing many of their plants. Now he prepares the soil by drilling holes down three feet and filling them with sand before the plant goes in. Lavender by the Bay runs on

"In France, everyone grows lavender.
It is good for the body, for the mind and it's good for us."

— Serge Rozenbaum, Lavender by the Bay

computer-controlled irrigation in the summer. They farm with no pesticides or fungicides, controlling weeds by laying down black tarps over the soil.

The Rozenbaum's sell lavender products in their farm shop. From dried lavender bunches and culinary lavender buds to soap and sachets, visitors can sample the products and experience the many uses of lavender firsthand. This also includes Serge's lavender honey from bees kept on the property. "The good thing about lavender is you harvest in a week or two, but then you dry it out and it lasts forever," says Serge.

In addition to their farm shop located on the North Fork of Long Island in East Marion, they sell their lavender plants at many of the Greenmarkets in New York City.

"I wish I could bring that donkey with lavender bunches on the sides to New York City," Serge says. "Wouldn't you like to see that on the streets of Manhattan?"

http://www.lavenderbythebay.com

MAGIC OF LAVENDER · MAGIC OF LAVENDER

"He stung me!" Tears ran down her plump cheeks. "I petted him and he stung me." Betrayal glinted in her eyes.

This can happen in a field of lavender, although bees are usually too lavender-drunk to bother.

"Do you have a first aid kit?" the mother asked.

"I have one in the barn," I said. "But I have a magic potion right here." I waved the bottle of lavender oil in front of the young girl's curious eyes. "What's your name?"

"Michelle"

"Michelle, ma Belle." Yes, I did sing that, but it seemed to get her attention. I kneeled and took her hand in mine. "It's a magic flower potion." I doused the stung red hand with the lavender oil. "It takes all the pain away." The calming fragrance reached my nose.

She sucked in halting breaths and broke into a radiant smile. Apparently the magic potion had done the trick. –R.R.

THE HISTORY OF LAVENDER

Lavender was first domesticated in Arabia and brought to France around 600 BC. It was called Nardus by the Greeks and Spikenard by the people of India, but the Romans called it Lavender, from the Latin word 'lavare' which means to wash or 'livendula' which means bluish.

The ancient Egyptians may have used lavender for embalming and mummification and as a perfume to anoint their heads. The Greeks preferred to rub lavender oil on their feet and legs so the scent would envelop the body and ascend to the nose. Or they would rub lavender oil on the breasts since they were home to the heart. The Romans learned of lavender's healing and antiseptic qualities and incorporated it into their bathing rituals, and also used it to launder clothing and linens. Lavender is mentioned in the Bible as Spikenard or Nard, used for healing and soothing.

The healing powers of lavender were recorded in 77 AD by the Greek military physician, Dioscorides, who served under the Roman emperor Nero. He wrote that lavender relieved indigestion, headaches, and sore throats. Lavender oil was used to treat skin ailments and clean wounds and burns. Another Greek writer, Pliny the Elder, cited the benefits of lavender to help with menstrual cramps, upset stomachs, kidney disorders, jaundice, and insect bites. Roman soldiers took lavender oil into the battlefields to dress wounds and aid healing. Roman households used lavender buds on their floors to freshen the air, in their beds to encourage passion, and in their baths and on their bodies as a perfume.

Then took Mary a pound of ointment of spikenard,
very costly, and anointed the feet of Jesus, and wiped his feet with her hair:
and the house was filled with the odour of the ointment. John 12:3, King James Bible

English women sewed lavender into sweet bags to freshen the air and linens and repel insects. They mixed the buds with beeswax to polish furniture. They dried laundry on lavender bushes planted near the laundry room. Queen Elisabeth wore lavender perfume and drank lavender tea to soothe her migraines.

In France, Henrietta Marie, wife of King Charles I, bathed with lavender soap and put lavender buds in bowls for potpourri. King Charles VI had lavender sewn into his seat cushions. In the 16th Century, the French recognized lavender's protection against infection, noting that many glove makers using lavender to scent their gloves escaped cholera.

In the 17th Century, Europeans turned to lavender to protect themselves from the Great Plague. Street vendors sold bags of lavender, lavender oil, and Four Thieves Vinegar, used by grave robbers to wash plague victims' belongings.

Queen Victoria led English ladies full force into the lavender crusade. They hung muslin bags of lavender in their wardrobes and stuffed them in drawers and bed sheets. Lavender was used to repel insects, lice and fleas. They used it in household soap and furniture polish and in potpourri. Small bags were tucked into their corsets to attract suitors. England became known world-wide for its lavender products.

In 1910, French chemist and one of the founders of modern aromatherapy, Rene Gattefosse, survived an explosion in the laboratory but his arm was badly burned. He treated his arm with lavender oil which eased the pain and healed the burn quickly with no infection or scarring. Lavender oil was used in the World War I to dress wounds, since antiseptics were in short supply.

English Quakers brought lavender to America and grew it commercially in herb farms and sold their herbal concoctions to the public.

Today the largest producer of lavender is France. The heady herbal flower has found its way around the globe, grown commercially in Great Britain, Belgium, Bulgaria, Croatia, Germany, Spain, Tasmania, Netherlands, Russia, Japan, Canada, China, and North America. Lavender's transforming presence in the United States has inspired this book.

"There's flowers for you;

Hot lavender, mints, savory, marjoram;

The marigold, that goes to bed wi' the sun,

And with him rises weeping; thes are flower

Of middle summer, and I thek they are given

To men of middle age."

–William Shakespeare *The Winter's Tale*

www.ingramcontent.com/pod-product-compliance
Lightning Source LLC
Chambersburg PA
CBHW041100210326

41597CB00005B/143